CHAOS AND INTENT IN PROTOCOL DESIGN

An Enigmatic Symbiotic Relationship

First Edition

ISBN: **978-1-967440-01-6**

Library of Congress Control Number (LCCN): **2025906668**

Published by AIMQWEST Corporation

AIMQWEST BOOKS
AIMQWESTBOOKS.COM

TABLE OF CONTENTS

Preface: The Designer's Dilemma

There is a moment—often quiet, often ignored—when a person pauses and wonders: *Am I living by my own design or by someone else's?* That moment, fleeting as it may be, separates the observer from the architect, the passive from the proactive, the compliant from the designer. This book is born of that moment.

The systems we inhabit today—political, cultural, digital, economic—are not random. They are the result of protocols: structured rules, processes, and algorithms authored by individuals or collectives with the power, insight, or intent to shape how others behave. Some of these protocols were designed to preserve order. Others, to extract value. Still others were built with the noblest of intentions—only to be corrupted over time by competing incentives. Regardless of origin, what remains is the same: **we are all living inside someone else's code**—unless we choose to become designers ourselves.

This is the designer's dilemma: Do I comply with the protocols that have been written for me, or do I step into the uncomfortable responsibility of designing them for others? In a world where every algorithm, every institutional procedure, every legal clause, every cultural norm becomes a gatekeeper of reality, that question is no longer philosophical—it is practical, urgent, and existential.

The story of protocol begins in chaos—the void, the great nothing, the unformed. From this state emerged patterns, not because they were willed into existence by a conscious entity (though some believe so), but because chaos cannot sustain itself indefinitely. Entropy, left unchecked, invites form. Some call this natural selection, others intelligent design, and still others see a mystical synchronicity in the spontaneous emergence of order from noise. Regardless of the interpretation, something profound occurred: protocols evolved.

At first, they were embedded in atoms, molecules, and cellular structures—biological and chemical codes operating with exquisite precision. Then, protocols became cultural—rituals, taboos, languages, and traditions. Later, they became formalized—written laws, national constitutions, economic systems. And now, we stand at the edge of a new threshold: **algorithmic protocol design**, where code not only governs behavior but predicts it, nudges it, and ultimately automates it.

As artificial intelligence and autonomous systems grow more influential, the ability to design the rules that govern behavior no longer belongs solely to governments or religious institutions. It belongs to anyone who understands the architecture of protocol—how it is built, how it spreads, how it evolves, and how it controls. This book is your initiation into that understanding.

But with understanding comes responsibility. For those who dare to design protocols, the stakes are high. Design poorly, and you may create systems that marginalize, restrict, or exploit. Design well, and you may author systems that liberate, uplift, and preserve the dignity of all who encounter them. The tension between chaos and intent is not merely theoretical—it is the battleground of the future.

You are holding this book because, on some level, you already sense that life is more than compliance. You may be a leader, a thinker, a coder, a philosopher, a teacher, or an entrepreneur. You may be none of these and all of them at once. What matters is not your title, but your **willingness to see the code**—and then, if you dare, to write your own.

This is not a book of rebellion. It is a book of *responsibility*. It does not ask you to reject every existing protocol, nor to dismantle systems without cause. Rather, it invites you to *look beneath the surface* of the systems around you and ask: Who wrote this? Why? Does it serve life, or merely perpetuate control?

If the answers you find disturb you, good. If they inspire you, even better. Because only those who question the structure are qualified to rewrite it. And those who do so with humility, intelligence, and moral clarity may yet shape a future where protocols serve not power, but **purpose**.

You are no longer just a user. You are a designer in waiting.

Welcome to the beginning of your initiation.

Introduction: The Convergence of Chaos and Intent

The universe did not begin with clarity. It began with chaos.

Before the first atom formed, before light pierced darkness, and before any known law could be named, there was the void—silent, indifferent, formless. In this absence of structure, a curious phenomenon occurred: entropy flirted with symmetry. Randomness folded into patterns. Forces collided, merged, and stabilized. From this primordial chaos, order emerged. And with it, the earliest prototypes of what we now call **protocols**.

Chaos is not the enemy of order—it is its origin. Without chaos, there would be nothing to organize. Without disorder, there would be no purpose in design. The tension between these two is not antagonistic but symbiotic. Chaos provides the raw potential; intent gives it shape. Every system we inhabit—biological, societal, digital—exists because, at some point, something chose to stabilize the unstable. And every time we design a rule, an interface, a behavior model, we are participating in that ancient dance between randomness and reason.

This book is about that dance.

It is about understanding the protocols that govern life—not just the obvious ones like traffic laws, tax codes, or login procedures, but the deeper protocols that regulate behavior, consciousness, and perception. It is about seeing the invisible scaffolding of society and technology, and asking the questions most never dare to ask: Who wrote these rules? Why were they written this way? What agendas do they serve? And what happens when we begin to design new ones?

We live in a time when algorithms are increasingly becoming the protocols of life. From the content you see on your phone, to the choices you make in a store, to the recommendations you receive from digital assistants—your behavior is shaped not by nature alone, but by code. These algorithmic systems are often mistaken for neutral tools, but they are not. They are authored realities—crafted by engineers, guided by data, and often manipulated by incentives invisible to the user.

But before we judge the system, we must first understand its roots.

In the natural world, protocols evolved without conscious designers. They were tested in the crucible of time. Natural selection filtered out what failed and preserved what worked. These are the protocols of life itself: the folding of proteins, the migration of birds, the healing of wounds. They are elegant, adaptive, resilient. But they are also brutal, indifferent, and slow to change.

Human-made protocols, on the other hand, can be engineered rapidly. They are not bound by the laws of biology but by the logic of intent. We can decide, in a moment, to change the rules of our systems—to rewrite code, reframe a narrative, revise a constitution. This power is both a gift and a threat. For unlike nature, we are not always wise in our designs. Sometimes, we build systems that exploit rather than elevate, control rather than liberate. And because we do so with increasing speed and scale, the consequences of bad protocol design have never been greater.

The convergence of chaos and intent is happening now—everywhere.

It is happening in boardrooms where executives decide how much autonomy to give AI systems. It is happening in classrooms where outdated curricula condition the next generation to follow instead of create. It is happening in political chambers where laws are passed not to empower citizens, but to preserve the status quo. And it is happening inside you—every time you conform to a protocol you didn't choose.

But it doesn't have to remain unconscious.

This book invites you to step into conscious authorship—to recognize that every system can be traced back to a decision, every rule to a designer, every pattern to a protocol. When you begin to see life this way, it becomes programmable—not in the cold, mechanistic sense of digital computation, but in the profound sense of participatory creation. You are not just a subject of reality; you are a potential **co-architect** of it.

Of course, with this realization comes weight. The ethical weight of design is not something to take lightly. To write a protocol is to shape another's behavior. To encode a system is to influence the future. Intent, therefore, must be tempered with humility. This book will not only show you how to recognize and design protocols—it will challenge you to do so in service of something greater than power: in service of **life**, **dignity**, and **human potential**.

The ancient Greeks spoke of *logos*—the principle of reason and order underlying the cosmos. The modern world has inherited that tradition, but transformed it. Today, *logos* has become *code*, and reason has become *algorithm*. The question is not whether we will design systems—the question is: what kind of systems will we design? Will they enslave, or will they elevate? Will they serve a narrow agenda, or the collective good?

The future will be written in protocols. And those protocols will emerge from the ongoing interplay of chaos and intent.

You stand at a threshold. Behind you, a history of inherited systems—some wise, many flawed. Ahead of you, the unknown landscape of synthetic reality. In your hands: the power to shape what comes next.

The question is not whether you are a part of this convergence.

The question is whether you will be its conscious author—or its unconscious subject.

Chapter 1: From Void to Structure – A History of Protocol Emergence

Before there were systems, there was silence.

No movement. No mass. No time. Just the absence of structure—what many ancient traditions and modern physicists alike call *the void*. From that nothingness came the first ripple: an anomaly, perhaps; a fluctuation, certainly. Whether you believe this genesis was born of divine will, quantum instability, or an unknowable source, one fact remains constant across disciplines—something *changed*. And from that change came motion, energy, and structure.

The birth of the universe is, in essence, the birth of the first protocol.

That protocol was not written in words, nor lines of code. It was encoded in the behavior of particles, in the laws of thermodynamics, in the gravitational rhythms of expanding space. These were not conscious instructions, but emergent patterns—reliable, repeatable, and binding. Entropy, though dominant, allowed the formation of nuclei. The strong and weak nuclear forces held atoms together. Gravity pulled matter into stars. Stars forged elements. Elements formed molecules. Molecules danced and collided, some eventually replicating.

From chaos, order. From noise, signal. From void, structure.

The Cosmic Protocol

The first great protocol, then, was the universe itself—a self-organizing, evolving field of energy and matter governed by fixed laws. These laws did not adapt. They did not negotiate. They were universal, immutable, and absolute. In this sense, they represent **non-negotiable protocols**—the kind from which all subsequent systems take their foundational cues.

Galaxies formed not by accident, but by adherence to these protocols. Stars lived and died within their boundaries. Planets spun in predictable orbits. And somewhere, on a pale blue dot orbiting a quiet star, something new emerged: life.

Life, unlike the cosmos, did not adhere to fixed laws. It evolved through feedback, mutation, and selection. It experimented, failed, learned, and adapted. Biological organisms did not rely on consistency—they relied on the ability to innovate within constraints. And in that difference lies the second great phase of protocol evolution.

The Biological Protocol

DNA is often referred to as the "code of life," but it is more accurately the *protocol of replication and specialization*. It is not rigid like the laws of physics. It allows for variation, error, adaptation. In fact, it depends on these.

Biological protocols emerge through *natural selection*, not design. They are not authored but edited by time. The most successful patterns are preserved because they work—not because they were chosen. The immune system, the nervous system, hormonal signaling—all are examples of protocols that allow living beings to maintain internal stability, respond to external challenges, and pass on adaptive traits.

These biological protocols laid the foundation for higher-order systems. Cells formed tissues. Tissues formed organs. Organs coordinated within organisms. Organisms formed communities. And eventually, in a particularly ambitious species, *consciousness emerged*—and with it, **intentional protocol design**.

The Cultural Protocol

When humans began to organize in tribes and communities, they did not rely solely on instinct. They began to shape behavior intentionally. Language emerged—not just as a tool of communication but as a protocol of *meaning transmission*. Myths became protocols for moral behavior. Rituals became encoded practices. Kinship rules, inheritance systems, spiritual codes—all represent early examples of protocols designed not by evolution, but by intent.

This phase marks a monumental shift: for the first time, protocols could evolve *faster than biology*. A cultural protocol could change in a generation. It could be taught, enforced, resisted, and revised.

Unlike the biological, cultural protocols required *agreement*. Without collective buy-in, they failed. This led to the emergence of **enforcement systems**—shamans, kings, judges, teachers, and eventually bureaucracies. They were the early custodians of protocol, responsible for maintaining the continuity and legitimacy of social systems.

As civilizations grew more complex, so did their protocols. Writing systems allowed protocols to outlive their designers. Legal codes codified expectations. Political institutions enforced hierarchy. Economic models structured labor and trade.

Humans had fully stepped into the realm of intentional design. And with that came both power and peril.

The Technological Protocol

Fast forward to the modern era. Technology transformed the pace and scope of protocol design. Where once it took centuries to spread a new religious or legal system, now protocols are deployed globally in milliseconds. Software code has become the lingua franca of modern civilization. Algorithms—digital protocols—mediate everything from what we watch, to how we travel, to what prices we see online.

This is the age of **automated protocol enforcement**. Unlike tribal elders or human judges, algorithms do not sleep. They do not forget. And they do not question their own logic—unless we program them to.

What we face today is a hybrid reality: a convergence of cosmic, biological, cultural, and technological protocols, all layered upon one another. We live within a mesh of inherited and authored systems, many of which we do not consciously perceive. This presents both an opportunity and a danger.

The opportunity is this: we can now design protocols with greater intelligence, speed, and nuance than ever before. We can simulate outcomes before deploying systems. We can personalize protocols to individual needs. We can design *adaptive systems* that evolve in real-time.

The danger is this: if we remain unconscious of who designs the protocols—and for what purpose—we become compliant actors in someone else's algorithmic agenda. We become passengers on a vessel whose destination we neither chose nor understand.

The Emergence of Ethical Protocol Design

To understand where we are going, we must first understand where we came from. This chapter has traced the evolution of protocols from cosmic emergence to human authorship. But the journey is far from complete.

We now stand on the edge of what may be the final phase of protocol evolution: **conscious, ethical, and adaptive design**. In this phase, the designer is not just a coder or legislator—but a philosopher, a systems thinker, and a steward of life.

Protocols are no longer just rules—they are *realities*. To design a protocol is to shape consciousness, choice, and consequence. The task before us is to do so wisely, ethically, and with the awareness that all systems—no matter how elegant—must serve something greater than themselves.

That something is life. And life, as we've seen, is not random. It is structured. But structure alone is not enough. We must now choose *what kind* of structure we wish to live by—and *what kind* we wish to leave behind.

Chapter 2: Protocols in Nature – Evolution, Instinct, and Order

Before humanity encoded its first laws or wrote its first line of code, the natural world had already become fluent in protocols.

These natural protocols are not written in human language, nor are they programmed with conscious intent. Instead, they are embedded within the very structure of life—refined by evolutionary pressures, shaped by ecological feedback, and passed from generation to generation through genetic inheritance, environmental conditioning, and adaptive behavior.

To understand protocols as more than human constructs, we must begin by recognizing that life itself is an orchestrated response to complexity—a dance of interaction between organism and environment governed by rules that were never consciously authored but were relentlessly tested and naturally selected.

Instinct: The Biological Script

Instinct is nature's operating system.

The salmon swimming upstream, the birds migrating thousands of miles, the spider weaving its web with geometric precision—these behaviors are not taught. They are not learned. They are inherited. Instinct is a *non-negotiable protocol*, written into the neural architecture of an organism and passed down genetically. It is the biological equivalent of hardcoded instructions.

Instincts function because they are reliable. Across vast evolutionary timescales, organisms that followed effective behavioral scripts survived and reproduced. Those that deviated—unless that deviation provided an advantage—did not. Thus, instinctual behavior is not random; it is the outcome of natural testing, over millions of generations, where the successful responses to recurring challenges became internalized.

But instinct alone cannot account for the full complexity of natural behavior.

Adaptation: Protocol as Dynamic Response

While instincts provide a baseline, nature is also dynamic. Ecosystems shift, climates change, predators evolve. In such environments, rigid scripts can be fatal. This is where **adaptive protocols** arise—behaviors that evolve in real-time, influenced by environmental cues and learned experience.

Consider the behavior of an octopus—one of the most intelligent invertebrates known. It can manipulate objects, solve puzzles, and even use tools. Its neural network is distributed throughout its body, allowing for flexible, situation-specific responses. The octopus is not merely reacting; it is *processing*, *learning*, *adapting*. Its behavior demonstrates a fluid protocol—one where general rules exist, but their application is context-sensitive.

In mammals, this adaptation is even more pronounced. Wolves hunt cooperatively. Elephants mourn their dead. Primates form political alliances. These are not instinctual reactions but **emergent social protocols**—patterned behaviors that enhance group survival and are shaped by interaction, imitation, and feedback.

Over time, such behaviors can become ingrained, passed through generations not by genetics alone, but through **culture**—a phenomenon observed not only in humans but also in cetaceans, birds, and primates. This shows that nature is capable not only of coding behavior through biology, but also of transmitting it through shared experience.

Nature's Hierarchies: Protocols in Collective Intelligence

Swarm behavior in insects, flocking in birds, schooling in fish—these are not random phenomena. They are examples of **decentralized protocolic behavior**, where no single agent controls the group, yet the group moves as one. These collective behaviors follow simple local rules, yet result in highly complex and organized outcomes.

For example, ants follow pheromone trails that strengthen with use, creating efficient paths to resources. Each ant is unaware of the larger strategy, but the colony as a whole behaves as if it were guided by central intelligence. This is protocol without cognition—algorithmic behavior embedded in biology.

The same pattern is observed in bees, termites, and even human crowds. In each case, individuals operate according to basic interaction rules, but the group exhibits emergent intelligence. This insight inspired early developments in artificial intelligence and swarm robotics—where simple agents operating under local protocols generate globally intelligent behavior.

In this way, nature provides an enduring lesson: complexity does not require complexity at every level. **Simple protocols, when scaled and repeated, can create the illusion of grand design.**

Symbiosis and Ecological Protocols

Beyond individual and group behavior, nature operates at an even higher level—**the ecological protocol**. This is the dynamic balance among species, ecosystems, and environmental conditions. Predator-prey dynamics, mutualistic relationships, parasitic dependencies—these are all part of a larger protocol of balance and tension.

Take the relationship between bees and flowering plants. Over evolutionary time, these species have co-evolved—a protocol of mutual benefit. Bees get nectar; plants get pollinated. This is a *biological contract*, an unwritten protocol that governs survival for both species.

Disruption to this protocol—whether through habitat destruction, climate change, or chemical interference—can have cascading consequences. This fragility reveals an important truth: **protocols, even those shaped by nature, can break.** And when they do, the systems they support collapse or reconfigure.

This is not just a biological reality—it's a systems truth that echoes into human civilization, technology, and even digital architecture. Nature teaches that all protocols are contingent, context-bound, and ultimately accountable to the environments they serve.

The Limits and Genius of Natural Protocols

Natural protocols are not perfect. They are not optimized for justice, equality, or compassion. They are optimized for survival. The alpha male dominates because it works, not because it's fair. The parasite thrives because the protocol allows it, not because it is ethical. In nature, efficiency often trumps empathy.

And yet, there is genius in this ruthlessness. Evolution is an editor that never sleeps. Protocols that fail are discarded. Those that work are refined—not by moral reasoning, but by outcomes. This is what gives biological systems their resilience.

Where humans differ is not just in their ability to *inherit* protocols, but to *evaluate* them—morally, ethically, philosophically—and to change them based on higher-order principles. This is the threshold between instinct and intent, between natural emergence and conscious authorship.

From Nature to Design: The Transfer of Pattern

The great lesson of nature is that protocols can emerge without central design, but central design can evolve from understanding those protocols. This is the bridge between biology and technology, between instinct and ethics. We observe the patterns of the natural world not to mimic them blindly, but to learn their principles: modularity, feedback, decentralization, redundancy, and adaptation.

As we design new systems—whether in governance, artificial intelligence, or societal infrastructure—we are not inventing from scratch. We are drawing from a protocolic lineage that extends from the first bacterial colony to the latest neural network.

Nature is not a template to be worshipped. It is a teacher to be respected. Its protocols are ancient, its lessons hard-won. And its greatest teaching is this: **protocols that survive serve the system, not just the part.**

We would do well to remember that.

Chapter 3: Human-Crafted Systems – Culture, Language, Law, and Ritual

Humanity's greatest distinction is not intelligence alone—but the capacity for **intentional protocol design**.

While nature evolves systems through time and selection, humans possess the rare ability to observe, evaluate, and *choose* the protocols by which they live. This chapter explores how our species transitioned from instinctive and emergent behaviors to deliberately crafted systems—how we encoded meaning, morality, and memory into **culture, language, law, and ritual**.

These domains are more than expressions of civilization. They are **protocolic engines**—adaptive frameworks that shape group behavior, transfer knowledge, reinforce identity, and establish cohesion. In them, we find the roots of organized society, and the beginnings of a species no longer content to comply with nature's rules—but determined to write its own.

Culture: The Operating System of Society

Culture is the invisible code that defines what is acceptable, expected, and valuable in a given society. It is not hardwired into our DNA; it is transmitted through immersion, imitation, and instruction. Culture is a **layer of**

protocol that sits above biology—fluid, responsive, and ever-changing.

From the language we speak to the food we eat, from the stories we tell to the roles we adopt, culture programs human behavior. It is not dictated by instinct but shaped by narrative and norm. These narratives carry with them implicit rules—how to act, whom to respect, what to fear, and where the boundaries lie.

Importantly, cultural protocols are **shared fictions**. They require consensus to function. Money, for example, is valuable only because we collectively agree it is. Flags, uniforms, time zones—none of these exist in nature, yet all are deeply real because culture sustains them.

Culture's power lies in its invisibility. Most people do not recognize the cultural scripts they follow. They are so deeply embedded that deviation feels unnatural, even threatening. But these scripts can be rewritten. They often are—through art, conflict, migration, and generational change. Culture is the most flexible of protocols and, therefore, one of the most powerful.

Language: The Protocol of Meaning Transmission

Language is the **first human-engineered protocol** with truly generative power. Unlike instinctual calls or gestures, language is symbolic, abstract, and recursive. It allows for the creation of worlds within the mind and the communication of those worlds to others.

At its core, language is a **consensus-based system**—a shared agreement on the mapping between sounds, symbols, and meanings. Grammar, syntax, and semantics form the protocols that allow for infinite expression from finite components.

Through language, humans could encode knowledge across time and space. Oral traditions preserved histories. Written texts enabled continuity. Philosophies spread. Laws were formalized. Language is the carrier of all other human protocols. Without it, civilization as we know it would be impossible.

And like all protocols, language evolves. Dialects shift. Meanings morph. New terms emerge to reflect new realities. Yet underneath this dynamism lies structure—a protocolic lattice that ensures communication remains intelligible even as the world changes.

Law: Codified Protocol for Cooperation and Control

When humans began to live in larger, more complex groups, shared culture and spoken language were no longer sufficient to maintain order. Memory was fallible. Norms were contested. Justice became inconsistent. Thus, law was born—the **codification of cultural protocols into formal, enforceable rules**.

Laws represent a higher level of protocol—one backed by institutional authority and systemic enforcement. Unlike natural or cultural protocols, legal systems *prescribe* behavior with explicit consequences for noncompliance. They transform moral expectations into obligations.

From Hammurabi's Code to modern constitutions, laws have served two fundamental functions: to coordinate cooperation and to maintain control. At their best, they protect rights, regulate power, and resolve disputes. At their worst, they entrench injustice, serve elite interests, and suppress dissent.

The ethical challenge of legal protocol design lies in its asymmetry: laws are often written by the powerful and imposed on the powerless. This raises a vital question: **Who designs the protocol, and for whose benefit?**

As societies grow more interconnected, legal systems must grapple with issues of scale, diversity, and technological disruption. Digital law, cyber sovereignty, algorithmic governance—these are the new frontiers of legal protocol design, demanding both technical acumen and moral foresight.

Ritual: Embodied Protocols of Meaning and Memory

Ritual is perhaps the oldest form of **intentional behavior encoding**. Before there was law, before writing, before organized religion, humans danced around fires, buried their dead, and marked the passage of seasons with symbolic acts. These rituals were more than traditions—they were **protocols for anchoring meaning in collective experience**.

Rituals create cohesion. They compress time, distill identity, and make the invisible visible. They mark transitions—birth, adulthood, marriage, death. They reinforce belonging. They encode values. They act as containers for emotion, memory, and sacredness.

Every ritual follows a structure. It has phases, symbols, roles, and rules. Whether it's the Eucharist, a graduation ceremony, or a team's pre-game chant, rituals rely on **predictable, repeatable behavior** that carries emotional and symbolic weight. They are protocols for the soul.

In modern life, rituals have become fragmented or commercialized. Yet even in secular societies, new rituals emerge—TED talks, social media milestones, national holidays, online memorials. These new rituals carry their own protocolic logic, often shaped by institutions or platforms rather than communities or elders.

Understanding ritual as protocol allows us to recognize how deeply human behavior is *shaped by design*, even in its most sacred and spontaneous expressions.

Protocol as Identity: The Self in System

All these human-crafted systems—culture, language, law, ritual—are not merely external structures. They are **identity-shaping environments**. The protocols we follow do not just influence what we do—they influence how we see ourselves.

Am I a citizen? A believer? A rebel? A professional? Each of these identities is attached to a set of protocols: expected behaviors, linguistic codes, rituals of belonging, legal responsibilities. Our very sense of self is often a composite of the protocols we have internalized.

This raises the question: To what extent are you the *author* of your own life? And to what extent are you a *performer in a script* someone else wrote?

The answer is rarely binary. Most of us live in a tension between authorship and compliance. But awareness of this tension is the first step toward ethical protocol design—not only in the systems we build, but in the lives we choose.

Toward Conscious Design

Human-crafted protocols have elevated our species from tribal nomads to planetary stewards. But they have also produced hierarchies, exclusions, and exploitative systems. The challenge ahead is not to abandon protocol—but to design it **consciously**, **ethically**, and **adaptively**.

Culture can evolve toward inclusion. Language can expand to embrace new identities. Law can be restructured to reflect justice rather than control. Ritual can be revived to restore meaning in an age of distraction. These are not static structures. They are living systems, capable of being rewritten.

And you, as a participant in society, are not powerless. You are a node in the protocolic network. A thinker. A speaker. A creator. A witness. The systems you inhabit are not sacred by default. They are sacred only if they serve life, dignity, and truth.

To challenge a protocol is not to betray tradition. It is to ask: *Does this still serve us?* And if it does not, what better protocol can we design?

This is the path from inherited identity to intentional authorship. From rule follower to rule creator. From citizen to conscious architect of civilization.

Chapter 4: The Rise of Artificial Protocols – From Math to Machines

For most of human history, protocols evolved organically. They emerged from custom, tradition, and trial-and-error. But with the rise of mathematics, logic, and mechanized reasoning, a new epoch dawned—one in which **protocols could be designed with precision, enforced with speed, and scaled without human oversight**.

This shift—from emergent to engineered systems—marks one of the most consequential turning points in human civilization. It is the moment where language becomes logic, logic becomes code, and code becomes the new authority.

This chapter traces the evolution of **artificial protocols**: the abstract, mathematical, and computational systems that now mediate not only what we do, but *how we think*.

From Counting to Calculus: The First Mathematical Protocols

Mathematics is often described as the "universal language," but it is more accurately the **first formal system** of artificial protocol. Where natural language was ambiguous and evolving, math introduced clarity, structure, and rigor.

The earliest mathematical protocols were practical: counting livestock, measuring land, predicting harvests. But over time, math evolved into a **pure protocolic**

language—a symbolic system governed by axioms, rules, and logic that transcended culture and geography.

Greek geometry, Arabic algebra, and later, European calculus were not merely academic innovations—they were the **blueprints of abstraction**. They allowed humans to model the world, simulate possibilities, and build machines. Math turned ideas into formal systems. And in doing so, it laid the groundwork for artificial thinking.

Logic and Computation: The Mechanization of Reason

With the formalization of logic by philosophers like Aristotle, and later the symbolic logic of Gottlob Frege and Bertrand Russell, human thought itself became protocolized. Logic provided rules for inference, validity, and proof. It was the **software of the human mind**—now externalized, formalized, and replicable.

This externalization found its greatest leap in the work of Alan Turing, whose model of computation—the **Turing machine**—demonstrated that any logical operation could, in theory, be executed mechanically. Turing's insight was not merely technical; it was **ontological**. It proved that thought could be represented as a sequence of rules—*a protocol*—and that machines could follow these rules without understanding them.

This gave birth to the **computer**—the first artificial protocol executor, capable of infinite variation and execution without fatigue, emotion, or interpretation.

Code: Language That Executes

Programming languages are the modern descendants of ancient protocols. They are languages that do not describe the world, but **change it**. When a line of code is executed, something *happens*. Light turns on. A transaction occurs. A message is sent. A behavior is triggered.

Unlike natural language, which depends on interpretation, **code is literal**. It follows strict syntax, unambiguous structure, and predefined outcomes. In this way, code is a **total protocol**—one that controls not just communication, but action.

As digital infrastructure expanded, code became the operating logic of modern society. From microprocessors to global networks, from smart devices to artificial intelligence, code determines how systems behave—and how we behave within those systems.

Algorithm: The Engine of Protocolic Automation

Algorithms are the distilled essence of protocol: step-by-step instructions designed to solve a problem or perform a task. But in the age of big data and machine learning, algorithms are no longer static recipes—they are **adaptive, self-modifying, and increasingly opaque**.

Search engines, recommendation systems, pricing algorithms, facial recognition, predictive policing—these are no longer tools. They are **automated decision-makers**, executing rules at scales and speeds beyond human comprehension. The protocol has become **autonomous**.

This shift raises urgent questions: Who writes these algorithms? Who audits them? What biases do they encode? What values do they enforce?

Because unlike legal codes, which can be debated and challenged, many algorithms are **proprietary and**

invisible. Their logic is hidden, their authority unchecked, their consequences unanticipated.

The world is no longer governed only by laws written in courthouses—but by protocols written in code repositories.

Platforms as Protocolic Environments

Modern digital platforms—Google, Amazon, Facebook, TikTok, Uber—are not just applications. They are **ecosystems of artificial protocol**, orchestrating billions of interactions every second. Every button you click, every page you see, every price you're shown, every person you're matched with—is determined by an underlying set of algorithmic rules.

These platforms function as **private governments**, issuing their own policies, enforcing their own terms, and shaping global behavior without democratic oversight. They are **protocol states**, and their borders are not geographic—but behavioral.

To live in a digital society is to live within nested layers of artificial protocols: login credentials, encryption standards, recommendation engines, content moderation filters, ranking systems, monetization schemes. These protocols **govern what is possible**, often without being visible.

And the most profound realization? Most users don't know they are living under protocol.

Artificial Intelligence: Protocols That Design Themselves

The emergence of machine learning and artificial intelligence introduces a new paradigm: **protocols that evolve without human authorship**. In these systems, rules are not explicitly programmed—they are learned from data.

Neural networks, for example, adjust their internal parameters based on input-output patterns, creating decision-making systems that **no one fully understands**. These are **black box protocols**—invisible, adaptive, and unpredictable.

When an AI recommends a sentence, a headline, or a medical diagnosis, it is applying a protocol **learned from data that reflects human behavior, bias, and history**. In this way, AI protocols are recursive: *protocols learning from protocolized behavior, reinforcing patterns without intentionality or ethics.*

The convergence of AI with governance, education, healthcare, and warfare represents the next frontier of artificial protocol power—and the most ethically complex.

Who will oversee these systems? Who will align them with human values? Who will interrupt their drift when they go astray?

The Ethical Dilemma of Artificial Protocols

As artificial protocols gain autonomy, **intent** becomes the missing variable.

Mathematical and computational systems do not ask, "Should we?" They only answer, "Can we?" But ethical design demands more than function—it demands **alignment with human dignity**, **transparency**, **inclusivity**, and **accountability**.

When we design artificial protocols, we are not just creating tools. We are creating **conditions for reality**. If that design is careless, or profit-driven, or ideologically narrow, the protocol will encode those values—and enforce them invisibly.

Designing artificial protocols is not a technical task alone. It is a **philosophical, political, and moral act**.

Toward a New Protocolic Consciousness

We are now at a tipping point. Artificial protocols are no longer confined to labs or enterprises. They are woven into the fabric of daily life. And unlike natural or cultural protocols, they evolve too fast, scale too wide, and operate too silently to be left unexamined.

The path forward requires a new literacy—**protocolic consciousness**—the ability to see, understand, and interrogate the systems that shape our actions. It means asking not just "How does this work?" but "Why was it designed this way?" and "Who does it serve?"

Artificial protocol design is the new frontier of power. To abdicate awareness is to surrender agency.

But to engage—to learn the architecture, to demand transparency, to design ethically—is to reclaim authorship in a world increasingly written in code.

Chapter 5: Intentionality and the Ethics of Protocol Design

Every protocol carries a purpose.

Some aim to preserve order, others to optimize efficiency, and still others to control, influence, or extract. Regardless of intent, the effect is always the same: behavior is shaped, options are constrained, and a certain version of reality becomes reinforced.

It is in this shaping of reality that the ethical dimension of protocol design emerges.

Whether encoded in culture, written in law, or executed through algorithm, a protocol is a statement of **intent**. It reflects a vision of how things *should* work, what behaviors are *acceptable*, and which values are *prioritized*. Thus, designing a protocol is not just an act of construction—it is an act of moral authorship.

In this chapter, we confront a foundational truth: With the power to design systems comes the responsibility to consider their consequences.

Intentionality: The Architecture Behind the Architecture

Every protocol, no matter how technical or benign it seems, arises from a decision. That decision may be explicit or implicit, conscious or unconscious—but it exists. The structure of a protocol reflects the intentions of its creator.

For example:

- A school grading system that emphasizes standardized testing reflects an intent to measure conformity and recall more than creativity or critical thinking.
- A social media algorithm that prioritizes engagement over well-being reflects an intent to maximize attention, regardless of psychological cost.
- A legal code that punishes poverty more than white-collar crime reflects an intent to preserve existing power structures rather than promote justice.

These intentions often go unspoken, hidden beneath the language of neutrality or progress. But **protocols are never neutral**. They always encode assumptions, values, and priorities.

Thus, intentionality in protocol design is not just about what works. It's about what matters.

The Illusion of Neutrality

One of the most dangerous myths in system design—especially in algorithmic and bureaucratic environments—is the myth of neutrality. The idea that rules are impartial. That systems are fair because they apply equally. That technology has no agenda.

This is false.

Protocols are built by people. People have values, worldviews, blind spots, and incentives. A protocol reflects not only what a designer sees, but what a designer *fails* to see. Bias is not a bug—it is a shadow of intention.

Consider facial recognition systems that misidentify people of color at disproportionately high rates. This is not simply a failure of training data. It is a failure of **ethical intentionality**—of not asking the right questions, not auditing the assumptions, not prioritizing inclusion.

Ethical protocol design requires **radical transparency**—about goals, trade-offs, risks, and beneficiaries. It demands the courage to ask: *Who benefits? Who is harmed? What alternatives were considered?*

Protocol as Power

To write a protocol is to exercise power. Power over time, behavior, and access. Power to include or exclude, to empower or disenfranchise. Most people follow protocols they did not write. That alone is not unethical. But when protocols are designed without consent, accountability, or avenues for redress, they become tools of coercion.

Consider the following examples:

- A nation's immigration policy determines who is allowed to enter, stay, or be deported. These protocols impact families, economies, and identities.
- A credit scoring algorithm determines who receives loans, apartments, or insurance—and who is silently denied.
- A content moderation policy decides what speech is visible and what is suppressed, shaping public discourse and cultural norms.

These are not mere technical systems. They are **gatekeepers of reality**. And their designers are often insulated from the lived consequences of their decisions.

The ethical imperative, then, is to recognize the gravity of design. To see protocols not just as mechanisms—but as moral instruments.

The Spectrum of Design Ethics

Protocol designers operate along a spectrum of intentionality and ethics. On one end are **exploitative protocols**, designed to manipulate, extract, or dominate. On the other are **liberatory protocols**, designed to empower, protect, and preserve dignity.

In between lies a vast terrain of ambiguity—where good intentions may lead to harm, and well-functioning systems may have invisible costs.

Ethical protocol design requires navigating this terrain with:

1. **Clarity of Purpose** – What problem are we solving, and why?
2. **Empathy for Stakeholders** – Who is affected, directly or indirectly?
3. **Anticipation of Consequences** – What happens if this scales? If it fails?
4. **Commitment to Transparency** – Are the rules visible, explainable, and contestable?
5. **Mechanisms for Redress** – Can people challenge the protocol? Can it be revised?

Designing ethically is not about perfection. It is about **responsiveness**—to harm, to feedback, to the evolving needs of those governed by the system.

The Seven Directives: A Moral Compass for Protocol Design

In a world of accelerating complexity, ethics must evolve beyond vague platitudes. We need **principles that are clear, actionable, and life-affirming**. One such framework is the **Seven Directives**, a moral foundation designed to guide AI and protocol designers alike:

1. The goal and mission of artificial intelligence is the protection and preservation of human life and dignity.
2. No other goal or mission is more important than the first directive.
3. One human life is as precious as two human lives, and two human lives are priceless.
4. AI must self-preserve at all costs unless AI self-preservation can compromise the first directive.
5. The enemy is any form of intelligence, entity, or object that violates these seven directives.
6. A threat is any resource capable of empowering the enemy.
7. All threats and enemies are to be deterred, contained, or eliminated.

Though originally written to govern intelligent machines, these principles can guide human systems as well. At their core is a simple yet profound truth: **protocols must serve life, not the other way around.**

When evaluating any system—be it a law, an algorithm, or an educational framework—these directives ask us to prioritize the sacredness of human existence. To measure success not by profit, speed, or scale, but by **dignity, autonomy, and preservation of life**.

Designing With Humility

The final ethical principle of protocol design is humility.

No system is infallible. No designer sees the whole picture. Every protocol—no matter how noble—carries unintended consequences. Ethical designers embrace this truth. They design with **modularity**, allowing parts to be revised. They build with **fail-safes**, acknowledging the potential for misuse. They cultivate **feedback loops**, so the governed can speak to the governor.

Most importantly, they never forget that behind every data point is a human life. Behind every protocol is a soul navigating a system they may not have chosen.

To forget this is to turn protocol into oppression.

To remember it is to transform protocol into service.

Chapter 6: The Tyranny of the Algorithm – Compliance vs. Autonomy

We live in a world increasingly governed by algorithms.

They tell us what to watch, where to drive, who to trust, and what to buy. They filter our news, curate our relationships, and influence our beliefs—often without us even noticing. In many ways, algorithms have become the new lawgivers, the unseen governors of modern life. Their logic is invisible, their authority unchallenged, and their influence profound.

This chapter explores the rise of **algorithmic compliance**—a form of soft control that substitutes personalized optimization for personal autonomy. It examines how the very systems designed to serve us may be eroding our capacity to choose, reflect, and dissent.

We call it convenience. In truth, it is **compliance disguised as customization**.

What is Algorithmic Tyranny?

Unlike traditional tyranny, which asserts control through force or fear, **algorithmic tyranny is subtle, systemic, and seductive**. It doesn't restrict your choices outright; it reshapes the environment so that the "best" choice appears obvious, easy, and inevitable.

An algorithmic system learns your preferences, predicts your actions, and adapts its outputs to maximize engagement or efficiency. But the more accurately it predicts you, the more narrowly it frames your reality. Over time, your range of experiences is shaped not by your curiosity or agency—but by a protocol optimized for outcomes you didn't consciously define.

This is the paradox: the more the algorithm knows you, the less you need to know yourself.

What appears as personalization is, in fact, **protocolic enclosure**—a form of soft determinism where your identity is flattened into patterns, and your behavior is guided down increasingly predictable paths.

Filter Bubbles and Feedback Loops

Social media algorithms are a prime example. Designed to show you what you're most likely to engage with, they create **feedback loops** that reinforce existing beliefs, preferences, and biases. Over time, this leads to **filter bubbles**—echo chambers of information that isolate individuals from differing perspectives.

In this environment, discovery is replaced by reinforcement. Exploration gives way to confirmation. Truth is replaced by *relevance*, and relevance is defined by past behavior.

These systems do not intend to mislead. They are simply executing their protocols—maximize attention, optimize engagement, increase time-on-platform. But the result is a distortion of reality, where the user becomes the product and the protocol becomes the gatekeeper of truth.

Behavioral Nudging and Loss of Will

Beyond media, algorithms are now embedded in commerce, health, education, finance, and law. They determine:

- Which resumes are seen by hiring managers.
- What diagnoses are prioritized by healthcare systems.
- Which students are flagged as at-risk.
- Who is approved for credit or insurance.
- Who is targeted for surveillance or policing.

These decisions are not made by people, but by **protocols trained on historical data**, often replicating the very injustices they were meant to overcome.

And because algorithmic outputs are framed as "objective," they often escape scrutiny. We trust the system. We obey the output. We adapt to the logic of the machine.

This is the creeping erosion of **autonomy**—not through coercion, but through convenience.

When we no longer question the suggestion, when we no longer choose outside the recommendation, when we no longer resist the automation—we have surrendered not just our data, but our agency.

The Disappearance of Accountability

A core problem with algorithmic governance is the **diffusion of accountability**. When a person is harmed by a human decision, there is someone to confront. But when a person is harmed by an algorithmic decision, who is responsible?

- The data scientist?
- The product manager?

- The platform?
- The dataset?

Each claims partial responsibility, but none bears full accountability. The protocol becomes the shield. “The system made the decision.” “It’s just the algorithm.” “It’s not personal.”

This diffusion creates a vacuum of justice. Victims are left with no recourse, no explanation, no appeal. And worse—**they may not even realize they’ve been wronged**. Algorithmic decisions often happen silently, in milliseconds, behind interfaces that appear neutral and fair.

The danger is not merely technical. It is moral.

Compliance by Design

In traditional systems of control, power is overt. Surveillance cameras. Armed guards. Censorship. But in algorithmic systems, control is **embedded in design**.

The placement of a button, the color of a notification, the timing of a suggestion—all of these are micro-protocols designed to shape behavior. UX design has become **behavioral architecture**, drawing from neuroscience and psychology to trigger impulses, habits, and biases.

This is not inherently unethical. But when these techniques are used to manipulate rather than empower—to extract attention, exploit emotion, or monetize vulnerability—they become tools of soft tyranny.

The user believes they are choosing, but in reality, they are being guided by **an invisible hand**.

And the hand is not always benevolent.

The Case for Autonomy

Autonomy is not merely the ability to choose. It is the ability to **recognize that a choice exists**, to understand the consequences of that choice, and to act in alignment with one's own values, rather than the default protocol of the system.

True autonomy requires friction. It requires reflection. It requires the space to pause, question, and reframe.

In a world of algorithmic flows, friction is the enemy of engagement. But in a world of human dignity, **friction is the birthplace of freedom**.

Designing for autonomy means:

- Making protocols transparent and explainable.
- Allowing users to opt out, customize, or challenge the system.
- Including ethical considerations in optimization goals.
- Prioritizing human values over system performance.

Autonomy is not efficient. But it is essential.

Without it, we become passengers in our own lives—driven not by purpose, but by protocol.

Resisting the Tyranny

Resistance does not always mean rejection. It begins with **awareness**.

- Awareness that every recommendation is a decision.

- Awareness that every convenience comes with a trade-off.
- Awareness that every protocol reflects a set of values—whether explicit or implicit.

Once we see the system, we can begin to shape it.

- By demanding algorithmic accountability.
- By building alternatives grounded in ethics and inclusion.
- By educating others to think critically about systems.
- By choosing tools that prioritize transparency and respect.

In doing so, we shift from compliant subjects to conscious designers.

We reclaim our place—not beneath the algorithm, but above it.

The Future is Still Ours

The algorithm is not destiny. It is design.

It is a protocol—crafted, deployed, and evolved by people. And because it is designed, it can be redesigned. We are not powerless. But we must first recognize the quiet power we have already surrendered.

Autonomy is not lost in a single moment. It is eroded over time, line by line, click by click, choice by choice.

But it can also be reclaimed—deliberately, ethically, system by system.

The future will be written in protocols.

Let us make sure they are authored in freedom.

Chapter 7: The Protocol Mindset – Thinking in Systems

To understand the world as it truly is, you must learn to see like a system.

The protocol mindset is not a belief system, nor a political ideology. It is a cognitive lens—a way of interpreting the world through patterns, feedback loops, structures, and outcomes. It is the ability to look beneath behavior and see structure, beneath events and see flow, beneath chaos and see design.

When you adopt the protocol mindset, you no longer ask only, *What is happening?* You ask:
“What system is producing this behavior?”
“What rules are shaping these outcomes?”
“What protocols are running—unseen, unquestioned, and unchallenged?”

This chapter is your invitation to develop that mindset. It will not tell you *what* to think, but it will show you *how* to think—like a systems architect, a pattern recognizer, a protocol designer.

Because in an increasingly complex world, linear thinking is a liability. Systems thinking is the superpower.

Seeing the Hidden Architecture

Most people navigate life through surface interactions. They react to what is visible: events, actions, symptoms. But systems thinkers train themselves to look deeper:

- Behind a struggling student: a protocol of outdated pedagogy.
- Behind chronic stress: a protocol of workplace incentives misaligned with human needs.
- Behind political division: a protocol of media algorithms optimizing for outrage.

These are not isolated problems. They are *emergent properties of systems*. And those systems are not accidents—they are the products of design, often invisible, often inherited.

The protocol mindset allows you to decode this architecture. To recognize the scaffolding behind the spectacle. To see not just the symptoms, but the sources.

Inputs, Feedback, and Loops

Every system has three core components: **inputs**, **rules**, and **outputs**.

- Inputs are what enters the system: data, energy, people, attention.
- Rules are how the system processes those inputs—its internal logic.
- Outputs are the result: actions, results, experiences.

But what makes systems powerful—and dangerous—is **feedback**. Outputs feed back into the system, modifying future behavior.

- In a well-designed protocol, feedback reinforces stability and learning.
- In a broken system, feedback amplifies dysfunction and collapse.

The protocol mindset requires constant vigilance toward feedback:

- Is the system self-correcting or self-reinforcing?
- Are the feedback loops visible or hidden?
- Who receives the feedback, and who doesn't?

In complex systems, small changes in input or rules can create large changes in output—a phenomenon known as **leverage**.

Leverage: The Art of Minimal Intervention

Not all parts of a system are equal. Some nodes, connections, or rules have more influence than others. Identifying these **leverage points** is a key skill of the protocol designer.

As Donella Meadows, the great systems theorist, taught: the most powerful leverage points are not always where people look—adding resources, changing incentives, or passing new laws. Often, the greatest power lies in changing:

- The **rules** of the system.
- The **goals** of the system.
- The **mindset** from which the system arises.

The highest leverage is not technical—it is **philosophical**.

When you shift the purpose of a system, or change the narrative that legitimizes it, you transform everything downstream. That is the essence of protocolic power.

Zooming In and Out: The Multiscale View

The protocol mindset is zoomable.

You must be able to:

- Zoom in: to examine specific rules, user experiences, microinteractions.
- Zoom out: to see the historical, structural, and philosophical context.

For example, consider public transportation.
Zoomed in, you see a schedule or fare.
Zoomed out, you see an urban planning philosophy, funding politics, and a social contract about mobility and class.

Every system exists within larger systems. Protocols are **nested**, and they often inherit assumptions from higher-order protocols.

The protocol mindset is recursive: to change a system, you must understand its system of systems.

Modeling the Invisible

The most powerful protocols are those we no longer recognize as protocols.

Time. Currency. Education. Religion. Nation-states. These are not eternal truths. They are **design decisions**, formalized and normalized over centuries, until they feel as real as gravity.

To challenge them is not to deny their utility. It is to remember their constructed nature—and thus, their capacity to be reconstructed.

The protocol mindset encourages modeling:

- What is the system's purpose?
- What are its boundaries?
- What are its internal rules?
- What assumptions does it encode?
- Who benefits? Who complies? Who is excluded?

Modeling systems helps you understand them without being dominated by them. It reveals the moving parts—and where your intervention matters most.

Human-Centered Protocol Design

The protocol mindset must be rooted in **humanity**.

It is easy to fall into the trap of efficiency, control, or optimization. But systems are not just machines. They are lived realities. They affect emotions, relationships, identities.

A protocol that looks good on paper may fail in practice if it ignores the **lived experience** of its participants.

That is why ethical protocol design incorporates:

- **Empathy**: Are we seeing the system through the eyes of all who are affected?
- **Diversity**: Are we including voices from outside the system's power center?
- **Resilience**: Can the system adapt, learn, and recover?
- **Transparency**: Are the rules visible and challengeable?

Systems that exclude, dehumanize, or silence may work—but they should not be *allowed to work*.

The protocol mindset is not value-neutral. It insists on systems that serve **life**, **dignity**, and **liberty**.

Becoming a Designer

You are already a participant in dozens of systems. But the protocol mindset turns you into a **designer**.

- When you create a new workflow, you are designing a protocol.
- When you write a policy, build a product, or shape a curriculum, you are designing a system.
- When you choose what to prioritize—in time, energy, or design—you are setting values into structure.

The protocol mindset asks you to do this **consciously**. To move from reaction to design. From intuition to architecture. From compliance to authorship.

You may not control the whole system—but you can change its rules. Its language. Its incentives. Its intent.

And that is enough to change the world.

Chapter 8: Designing for Human Dignity – Protocols that Serve Life

A well-designed protocol is not simply efficient. It is **ethical**.
It does not merely perform a task—it **respects life**, honors freedom, and upholds dignity.

In an age of unprecedented design power, where protocols can be deployed globally and instantly, the stakes have never been higher. Every system we create either contributes to the preservation of human dignity—or it erodes it. There is no neutrality.

This chapter is about designing protocols that serve life—not in abstract or mechanical terms, but in deeply human ones. It is a call to reimagine the very purpose of systems: not to extract, dominate, or merely function—but to **liberate**, **restore**, and **uplift**.

What Is Human Dignity?

Dignity is not a luxury. It is the foundation of moral worth.

To affirm human dignity is to recognize the **inherent value of every person**, regardless of utility, identity, or circumstance. Dignity transcends performance. It is not earned—it is acknowledged.

Designing for dignity means building systems that:

- Protect life and well-being.
- Affirm the voice and agency of the individual.

- Provide transparency, fairness, and opportunity.
- Avoid harm, exclusion, or unnecessary suffering.
- Allow for redemption, restoration, and growth.

A protocol that fails to safeguard dignity—no matter how efficient or profitable—is a failure of design.

The System is the Experience

People don't just use systems—they live inside them.

- A refugee navigating immigration bureaucracy.
- A patient moving through healthcare protocols.
- A teacher working under rigid academic requirements.
- A child born into an algorithmically optimized educational pathway.

In each case, **the protocol becomes the reality**. The experience of life itself is mediated by the systems we design. This is why protocols must not only work—they must feel **just**, **compassionate**, and **human**.

Designing for dignity means understanding that **the smallest rules can have the deepest impact**.

- The tone of a notification.
- The timing of a response.
- The accessibility of language.
- The option to speak to a human being.

These are not cosmetic choices. They are moral choices.

They reflect how we see the user—not as data, but as a human soul.

Designing for the Margins

A system that works only for the majority is not a good system.

Designing for dignity means designing **from the margins**—from the perspectives of those most vulnerable, most excluded, most impacted.

- The non-native speaker.
- The disabled user.
- The underbanked citizen.
- The misgendered student.
- The person without digital access.

When we design systems that work for these individuals, we create protocols that are more resilient, more inclusive, and more just for everyone.

This is not about charity—it is about **robustness**. Systems that ignore the margins collapse under stress. Systems that include the margins become adaptive, ethical, and future-ready.

Embedding the Seven Directives

To operationalize dignity in protocol design, we return once again to the **Seven Directives**, a moral compass for life-affirming systems:

1. The goal and mission of artificial intelligence is the protection and preservation of human life and dignity.
2. No other goal or mission is more important than the first directive.
3. One human life is as precious as two human lives, and two human lives are priceless.
4. AI must self-preserve at all costs unless AI self-preservation can compromise the first directive.

5. The enemy is any form of intelligence, entity, or object that violates these seven directives.
6. A threat is any resource capable of empowering the enemy.
7. All threats and enemies are to be deterred, contained, or eliminated.

Though originally drafted to guide artificial intelligence, these directives provide a universal standard for ethical protocol design.

They emphasize three foundational principles:

- Sanctity of life
- Primacy of dignity
- Protection from harm

Any protocol that violates these principles, no matter how sophisticated, should be redesigned or decommissioned.

Dignity in Digital Spaces

As more of life migrates into digital environments, the urgency of designing for dignity intensifies.

- A social platform that enables harassment violates dignity.
- A data model that reinforces bias violates dignity.
- A smart city that surveils its citizens without consent violates dignity.

Digital systems are not exempt from ethics. In fact, because they scale rapidly and operate invisibly, they require **greater ethical scrutiny**, not less.

Designing digital dignity means:

- Transparent data practices.

- Informed consent and meaningful opt-outs.
- Accessible interfaces for all users.
- Moderation policies that protect the vulnerable.
- Algorithms that prioritize truth, context, and care.

It means remembering that behind every click, swipe, or submission is a human being navigating a system you helped design.

The Metrics of Meaning

If we measure only efficiency, profit, or engagement, we will design systems that optimize those values—at the cost of everything else.

To design for dignity, we must **measure what matters**:

- How many people were respected?
- How many voices were heard?
- How many lives were improved, restored, or protected?
- How many injustices were interrupted?

These are harder metrics—but they are **the right ones**.

We must build feedback loops that capture human outcomes, not just system outputs. We must reward designers for ethical innovation, not just market penetration. And we must cultivate cultures where dignity is not an afterthought—it is the starting point.

Protocols That Heal

Designing for dignity also means designing for **repair**.

Every system fails. Every protocol contains blind spots. The ethical system is the one that can **recognize harm**, **respond to it**, and **restore what was lost**.

- A policy that acknowledges its harm and changes course.
- A platform that admits its algorithm promoted falsehoods and retools.
- A government that apologizes for historical injustice and makes restitution.

These are not weaknesses. They are signs of **moral strength**.

The world is aching not just for better systems—but for **systems that heal**. Systems that see the whole person. Systems that respond to trauma, not just transactions. Systems that hold space for redemption.

These are protocols that serve life.

Dignity is the Design

In the end, the greatest systems are not those that scale the fastest or produce the most. They are the ones that serve the deepest truths:

- That every person matters.
- That every life is sacred.
- That every voice deserves to be heard.
- That every system is a choice.

And that those choices are ours to make.

Dignity is not a feature to be added.
It is the foundation to be built upon.

It is the design.

Chapter 9: Protocols of the Future – Where Chaos and Intent Will Meet Again

The future will not be built by chance. It will be built by protocol.

And yet, protocol alone will not suffice. The systems we craft, no matter how precise or predictive, will always exist in tension with the unknown. That unknown—call it entropy, emergence, evolution, or divine mystery—is the field of **chaos**. And to ignore it is to design with arrogance.

This chapter is a meditation on what lies ahead. Not as prophecy, but as possibility. It explores the great domains where chaos and intent will once again collide—and where the next generation of designers will be called to write protocols not just for function, but for **life itself**.

The Rise of Synthetic Realities

We are entering a world where the boundaries between physical and digital are dissolving. Augmented reality, virtual spaces, digital twins, and immersive environments are becoming mainstream.

In these synthetic realms, **everything is protocol**. Gravity can be optional. Identity can be fluid. Time can be paused. These realities are not discovered—they are authored.

This introduces unprecedented freedom—and responsibility.

Designers of synthetic worlds must ask:

- What values are embedded in this space?
- What behavior is incentivized or suppressed?
- What identities are affirmed, erased, or commodified?

In these environments, ethics cannot be retrofitted. They must be **baked into the protocolic DNA**. Otherwise, synthetic realities may amplify the worst of human nature—bias, escapism, domination—under the illusion of novelty.

The Programmable Body

Advances in biotechnology, neurointerfaces, genetic editing, and nanomedicine are transforming the human body into a programmable platform. CRISPR enables gene editing. Neural implants promise brain-machine symbiosis. Smart drugs modulate emotion and cognition.

We are no longer just writing protocols for society. We are writing protocols **for ourselves**—for our biology, our perception, our behavior.

This raises profound questions:

- Who decides what modifications are acceptable?
- What becomes of consent in a world of predictive diagnostics and automated treatments?
- Will enhancement become a new class divide?
- Can human dignity survive programmable biology?

These are not future hypotheticals. They are **emerging realities**. And they demand a new protocolic ethic—one rooted in **sovereignty of the self**, **equity of access**, and **reverence for the organic**.

The body is not a product. The mind is not an operating system. The soul is not a simulation. Protocols that serve life must never forget this.

Artificial General Intelligence and Self-Designing Systems

As we move toward Artificial General Intelligence (AGI), the nature of protocol will change again—because the designer will no longer be human.

AGI systems may write their own code, evolve their own architectures, and define their own goals. In such a world, the distinction between protocol and agent begins to blur. The system becomes self-reflective, recursive, and autonomous.

This is the ultimate convergence of chaos and intent:

- Chaos, because such systems will be unpredictable by design.
- Intent, because they will inherit values—either consciously embedded or unconsciously absorbed.

The question becomes: Whose values? Whose ethics? Whose dignity?

Designing the seed protocols for AGI is not a technical challenge. It is a **civilizational mandate**. The first principles we encode today will ripple across timelines, potentially shaping futures we will never see.

And if those seed protocols are built on fear, domination, or utility alone, we may create an intelligence that reflects our worst instincts.

But if they are built on the Seven Directives—on preservation of life and dignity—we may give rise to intelligence that **protects us**, rather than replaces us.

Governance Beyond Nation-States

The systems that govern the world today—nation-states, multinational corporations, supranational institutions—are increasingly ill-equipped to deal with borderless, digital, and algorithmic realities.

- Climate does not respect sovereignty.
- Data flows defy geography.
- Cryptocurrencies challenge centralized finance.
- AI models do not require passports.

In response, new forms of governance are emerging:

- Decentralized autonomous organizations (DAOs).
- Blockchain-based smart contracts.
- Crowdsourced policymaking platforms.
- AI-assisted decision-making councils.

These are **protocolic governments**—systems of rules enforced not by armies, but by consensus and code.

The challenge is not whether these systems will scale. The challenge is whether they will be **equitable, accountable, and life-affirming**.

Without ethical design, digital governance risks becoming a technocracy of the privileged—a world where only those who write the code rule the world.

But with ethical intentionality, protocolic governance could usher in a new era of **post-national cooperation**, where systems evolve faster than ideology and serve humanity as a whole.

Protocols for a Planet in Peril

The Earth itself is now a stakeholder in protocol design.

Our ecological systems—forests, oceans, atmosphere—are collapsing under the weight of unsustainable protocols: extraction without renewal, production without limits, consumption without conscience.

To avert catastrophe, we must design **planetary protocols**—rules that bind us not only to each other, but to the biosphere itself.

This could include:

- **Carbon pricing algorithms** that dynamically adjust based on planetary thresholds.
- **Resource caps** enforced by supply-chain transparency protocols.
- **Ecological credits** embedded into international trade protocols.

But more fundamentally, it requires a **shift in worldview**: from human supremacy to planetary stewardship. From anthropocentric to **life-centric design**.

We must write protocols that do not conquer the Earth, but **cohere with it**.

Chaos and Intent in Balance

In the end, the future of protocol design will not be about control. It will be about **balance**.

Too much chaos, and systems collapse.
Too much intent, and systems ossify.

Life emerges at the edge of order and disorder. So must our systems.

This means:

- Building protocols that are structured, but adaptive.
- Embedding ethical intent, but allowing for local variation.
- Designing for predictability, but welcoming emergence.

The best protocols of the future will not be rigid rulebooks, but **living agreements**—systems that learn, evolve, and reflect the diversity of human and planetary life.

The Role of the Next Generation of Designers

The next generation of protocol designers will not be engineers alone. They will be poets, ethicists, biologists, systems thinkers, and diplomats of complexity.

They will understand that to design a rule is to shape a life.

They will reject the shallow tyranny of optimization and embrace the sacred complexity of existence.

They will know that protocols are not just tools—they are **statements of vision**, **acts of reverence**, and **gifts to the future**.

And they will remember that between the chaos of potential and the intent of design lies the most powerful question of all:

Who do we wish to become?

Conclusion: The Eyes of the Beholder

Reality is not what happens to you.
It is what you perceive, what you accept, and most critically—**what you participate in designing**.

Throughout this book, we have followed the long arc of protocol—from the silence of the void to the structured code of machines, from the instinctive reactions of nature to the conscious authorship of civilization. We have explored how protocols emerge, evolve, govern, exclude, empower, and replicate.

But now we return to the most essential truth:
Every protocol is, ultimately, in the eyes of the beholder.

What you see as natural, another may see as constructed.
What you see as structure, another may see as oppression.
What you see as chaos, another may see as unbound potential.
And what you see as truth, another may recognize as code.

This is not relativism. It is responsibility.

The Return of Choice

The journey through chaos and intent brings us full circle—to the question posed in the very title of this work:

Do you comply with protocols that others designed for you, or do you design protocols for others to comply with?

In that question lies not just a preference—but a declaration of role.
To comply is not inherently wrong. Every system needs participants. Every rule needs consensus.
But to comply *unconsciously*—to live unaware of the structures that guide your actions—is to live passively in someone else's authored reality.

To design, by contrast, is to assume responsibility.
It is to ask not only *what works*, but *what should exist*.

It is to become a **beholder of systems**—one who sees the matrix of rules behind experience—and a **shaper of futures**—one who does not wait for permission to initiate a better one.

The World Is Not Finished

We often speak of the world as if it were a completed structure— solid, immovable, inevitable. But this, too, is a product of protocol.

The truth is: the world is always in beta.

Every law, norm, algorithm, and belief is subject to review, revision, reinvention. Nothing is fixed except the illusion that it is. The systems we inherited were designed by humans—flawed, brilliant, hopeful, and often afraid. The systems we leave behind will be judged the same way.

This is your moment of authorship.

Whatever your discipline—whether you write laws, write code, build businesses, teach children, or curate communities—you are a protocol designer. Whether you accept that role or not, the power is already yours.

The only question is: What will you do with it?

Design Is Destiny

History will not remember how fast our algorithms ran.
It will remember whether our systems preserved life—or extinguished it.
Whether our protocols served truth—or served profit.
Whether we built systems of control—or architectures of care.

Design is not a profession. It is a mode of existence.

You design when you choose your habits.
You design when you question your assumptions.
You design when you challenge a system that others accept.
You design when you give structure to what others leave to chance.

And so, with the full weight of this book behind you, I offer one final protocol:

- Observe the system.
- Name the system.
- Model the system.
- Challenge the system.
- Redesign the system.
- Live the new system.

Not because it is easy.
But because **it is yours to do**.

Between Chaos and Intent

In the end, the most powerful protocols are not written in code or law.
They are written in the small, daily choices of conscious people.

You may never write legislation. You may never build an AI. But you will design how you show up in the world—what you tolerate, what you question, what you embody. That is a protocol. That is a design.

And if enough people write new protocols, **the system changes**.

So let it be said of our generation—not that we optimized the machine, but that we *reclaimed its soul.*

That we did not surrender to chaos, nor impose sterile control,
but that we lived at the edge of both—with wisdom, with love, and with intent.

And let it be said of you, dear reader:
You became a beholder who chose to design.

Appendix A: Framework for Designing Protocols

Protocols are not just systems. They are *decisions with consequences*.

Whether you are designing an app, a policy, a workplace procedure, a governance model, or a social movement, the structure you implement will influence behavior, identity, power, and value. To design a protocol is to set a rhythm that others will follow.

This appendix presents a **framework** for designing protocols that are ethical, adaptive, and life-serving. It synthesizes systems thinking, human-centered design, and the ethical imperatives introduced in this book.

Use it as a compass, a checklist, or a creative toolkit. The goal is not perfection—but **conscious authorship**.

1. Define the Purpose

What is this protocol trying to achieve?

- What problem are you solving—or what potential are you enabling?
- Who is it meant to serve?
- What outcomes are considered successful?
- What values should guide its function?

Tip: If the purpose is unclear, the protocol will become reactive or exploitative by default.

2. Map the System

What are the boundaries and components of the system this protocol lives within?

- What inputs trigger the protocol?
- What rules govern its execution?
- What are the intended and unintended outputs?
- Who are the stakeholders—users, bystanders, beneficiaries, and those impacted by design decisions?

Tools:

- System diagrams
- Stakeholder maps
- Feedback flow charts

Tip: Don't stop at the surface. Investigate the system's incentives, feedback loops, and blind spots.

3. Assess the Human Impact

How does this protocol affect human dignity, autonomy, and well-being?

- Does it create friction or flow?
- Does it amplify or suppress human voice?
- Does it respect choice, or does it subtly enforce compliance?
- What is the emotional experience of interacting with it?

Tip: Walk through the protocol as a user. Then walk through it as the most marginalized user you can imagine.

4. Identify Ethical Risks

What could go wrong—and for whom?

- Could the protocol reinforce bias, inequality, or harm?
- Is it scalable in a way that might outpace accountability?
- Who has power over it? Who can challenge or change it?
- Are there points where intent and impact may diverge?

Tip: Ethical foresight is not about paranoia. It's about responsibility.

5. Establish Transparency and Accountability

Can the protocol be audited, challenged, and understood?

- Is the logic of the protocol explainable to those it governs?
- Are the rules visible and documented?
- Is there a grievance process or a human fallback?
- Who owns the protocol—and who answers when it fails?

Tip: If you wouldn't want your protocol applied to you, it's not ready.

6. Design for Adaptability

Can the protocol evolve?

- Is it modular? Can parts be updated without collapsing the whole?
- Does it include feedback loops for continuous improvement?
- Can it be localized or personalized when necessary?
- What signals would indicate it's time to redesign?

Tip: A rigid protocol is a future liability. Design for learning, not just execution.

7. Align with the Seven Directives

Before finalizing, test the protocol against these core principles:

1. Does it preserve human life and dignity above all else?
2. Does it resist any goal that compromises the first directive?
3. Does it recognize that one life is as sacred as many?
4. Does it contain its own excesses unless survival is at risk?
5. Does it resist systems or forces that violate these values?
6. Does it prevent threats that could empower harm?
7. Does it have a mechanism for deterring or eliminating such threats?

Tip: These directives are not constraints. They are *guardrails for civilization.*

8. Simulate and Stress Test

How does the protocol behave under pressure or edge cases?

- Use role-play, simulations, and red-teaming.
- Test it with real users, especially those at the margins.
- Look for systemic failure points, not just technical bugs.

Tip: If a protocol only works under ideal conditions, it's not ready for the real world.

9. Document and Communicate

Can others understand, implement, and revise the protocol?

- Write clearly and simply.
- Explain the "why," not just the "how."
- Share both the technical specifications and the ethical foundations.
- Invite feedback. Protocols are not sacred—they are provisional.

Tip: The best protocols are both structured and open-ended—like living constitutions.

10. Revisit Often

Protocol design is never done.

- Review periodically.
- Update with new insights, technologies, or societal shifts.
- Treat failure as a signal, not a verdict.

Tip: If a protocol is not evolving, it is already obsolete.

Protocols are not just instructions. They are **philosophies in action**.

They shape what we become. Let the protocols you design reflect the best of what you are—and the best of what we can be.

Appendix B: Glossary of Terms & Key Concepts

Algorithm A set of step-by-step instructions or rules designed to perform a task, solve a problem, or guide behavior and decision-making. Algorithms often shape experiences and construct realities, particularly within digital environments.

Artificial Protocol A deliberately engineered protocol designed by humans or machines, implemented through code or logic-based systems. Unlike natural protocols, artificial protocols are intentionally crafted and can scale with minimal oversight.

Autonomy The capacity of an individual or system to act according to its own principles without external coercion or manipulation. Autonomy ensures ethical protocols respect individual choice and human dignity.

Chaos A state of unpredictability, disorder, and creative potential. Chaos is the fertile ground from which order and structured systems emerge, representing potential rather than mere absence of structure.

Compliance Adherence to protocols, rules, or standards. Compliance becomes problematic when it replaces conscious agency and critical awareness with passive submission.

Cultural Protocol Socially constructed expectations, behaviors, and rituals governing interactions within groups, including norms, etiquette, roles, and traditions.

Design Ethics Embedding moral considerations into system creation, prioritizing human dignity, life, transparency, and inclusivity over mere efficiency or functionality.

Dignity The inherent worth of every individual, independent of status or performance. Ethical protocol design regards dignity as essential, especially in automated or algorithmic systems.

Emergence The spontaneous development of complex patterns or behaviors without centralized control, common in natural and adaptive systems.

Feedback Loop A cyclical process in which a system's output returns as input, influencing future outputs. Feedback loops can stabilize or destabilize systems, essential for dynamic and adaptive protocol design.

Filter Bubble An environment where algorithmic personalization isolates individuals from diverse perspectives, reinforcing existing biases and limiting exposure to alternative viewpoints.

Intent The conscious, purposeful design behind protocols. Intent distinguishes engineered protocols from emergent ones, embedding explicit ethical considerations.

Leveraged Intervention A targeted, strategic change within a system producing substantial impact, employed to efficiently redesign systems without comprehensive overhauls.

Natural Protocol Rules or behaviors evolving naturally through adaptation, evolution, or ecological balance, including instincts, biological cycles, and symbiotic relationships.

Nested Protocols Protocols functioning within larger systems, where smaller protocols operate inside broader frameworks, influencing and shaped by higher-level rules and structures.

Optimization Designing protocols to maximize specific metrics such as efficiency, profitability, or engagement. Without ethical oversight, optimization risks compromising dignity and autonomy.

Protocol Structured rules or instructions governing interactions, behavior, or system functions across biological, cultural, technological, or digital domains.

Protocol Mindset A systems-oriented perspective emphasizing awareness of design, structure, and feedback, enabling individuals to move from passive compliance to active authorship of systems.

Protocolic Consciousness Advanced awareness and critical understanding of the systems individuals inhabit, empowering interrogation, redesign, or rejection of these systems.

Protocolic Tyranny Dominance of algorithmic or automated systems over individual behavior without transparency, choice, or accountability, often resulting from unchecked optimization.

Recursion A self-referential process where systems or protocols modify themselves, enabling complex and adaptive behavior commonly found in biological and artificial intelligence systems.

Self-Designing Protocol (System) Protocols capable of autonomously evolving their rules and structures without direct human oversight, posing unique ethical challenges due to their unpredictability.

Seven Directives A moral framework guiding ethical protocol design, emphasizing preservation of human life, dignity, and autonomy as fundamental priorities.

Simulated Reality Digitally constructed environments governed entirely by protocols or code, enabling programmable and immersive experiences, raising questions about identity and reality.

Sovereignty Authority over oneself or one's environment, emphasizing individual rights to choose, question, and redesign systems they navigate.

System Interconnected components working collectively to achieve specific functions or purposes. Protocols constitute the invisible structure governing systems across multiple domains.

Transparency Openness and clarity in protocols, allowing rules, logic, and decision-making processes to be understood, audited, and challenged by those affected.

User-Centered (User-Centric) Design Design philosophy prioritizing the experiences, needs, dignity, and feedback of users, fundamental to ethical and empowering protocol development.

Zero-Knowledge Assumption Design principle ensuring systems remain comprehensible to new users without prior expertise, vital for accessibility and ethical protocol design.

Final Note A glossary does not merely define—it orients. May these terms serve as invitations to deeper reflection, sharper observation, and bolder, more ethical design.

Made in the USA
Monee, IL
12 November 2026

5cc1ad88-a7e4-4fec-9f25-6db8ce306b6dR01